30-SECOND
THEORIES

30-SECOND
THEORIES

The 50 most thought-provoking
theories in science, each explained
in half a minute

Editor
Paul Parsons

Foreword
Martin Rees

Contributors
Jim Al-Khalili
Susan Blackmore
Michael Brooks
John Gribbin
Christian Jarrett
Robert Matthews
Bill McGuire
Mark Ridley

METRO BOOKS
NEW YORK

METRO BOOKS
New York

An Imprint of Sterling Publishing
387 Park Avenue South
New York, NY 10016

This book was conceived,
designed, and produced by
Ivy Press
210 High Street, Lewes,
East Sussex BN7 2NS, U.K.
www.ivy-group.co.uk

Creative Director **Peter Bridgewater**
Publisher **Jason Hook**
Editorial Director **Caroline Earle**
Art Director **Michael Whitehead**
Commissioning Editor **Nic Compton**
Designers **James Hollywell, Les Hunt**
Concept Design **Linda Becker**
Illustrations **Jon Raimes**
Glossaries & Profiles Text **Tom Jackson**
Picture Research **Lynda Marshall**

ISBN 978-1-4351-0956-8

For information about custom editions,
special sales, and premium and corporate
purchases, please contact Sterling Special
Sales at 800-805-5489 or specialsales@
sterlingpublishing.com.

Manufactured in China

4 6 8 10 9 7 5

www.sterlingpublishing.com

CONTENTS

FOREWORD
Martin Rees

Our world is getting ever more complex, more baffling. Some pessimists argue that scientific progress—or indeed society itself—will clog up because of "information overload." I don't think that's a serious worry. As science advances, more patterns and regularities are revealed in nature. These advances cut down the number of disconnected facts worth remembering. There's no need to record the fall of every apple, because, thanks to Isaac Newton, we understand how gravity pulls everything—whether apples or spacecraft—toward Earth.

The simplest building blocks of our world—atoms—behave in ways we can understand and calculate. And the laws and forces governing them are universal: Atoms behave the same way everywhere on Earth—indeed, they are the same, even in the remotest stars. We know these basic facts well enough to enable engineers to design all the mechanical artefacts of our modern world, from radios to rockets.

Our everyday environment is too complicated for its essence to be captured by a few formulae. But our perspective on our Earth has been transformed by great, unifying ideas. The concept of continental drift, for instance, helps us to fit together a whole raft of geological and ecological patterns across the globe. Charles Darwin's great insight—evolution by natural selection—reveals the overarching unity of the entire web of life on our planet. Whatever our personal lives may be like, our environment is neither chaotic nor anarchic. There are patterns in nature. There are even patterns in how we humans behave—in how cities grow, how epidemics spread, and how technologies, such as computer chips, develop. The more we understand the world, the less bewildering it becomes, and the more we're able to change it.

These laws or patterns are the great triumphs of science. To discover them has required dedicated talent—even genius in many cases. But to grasp their essence is not so difficult. We all appreciate music, even if we can't compose or perform it. Likewise, the ideas of science can be accessed and marveled at by everyone.

Science impinges more than ever on our lives. Many political issues—energy, health, environment, and so forth—have a scientific dimension. How science is applied matters to us all. The important choices shouldn't be made just by scientists; they should be the outcome of wider public debate. But in order for that to happen, we all need a "feel" for the key ideas of science. And, quite apart from their practical uses, these ideas should be part of our common culture. The great concepts of science can be conveyed briefly—maybe even in 30 seconds—using non-technical words and simple images. That's the aim of this book, and we should hope it succeeds.

The theory of everything

Unification theory, one example of which is string theory, attempts to explain how everything in the universe is connected. Great scientists have spent years trying to develop a "theory of everything"—this book explains that research in just 30 seconds (see page 50).

Testing, testing
Unlike the pet theories that we all like to make up, scientific theories are supported by cold hard evidence, usually in the form of carefully planned and controlled experiments.

INTRODUCTION
Paul Parsons

Everyone's got their own pet theory. I should know. During my time as editor of the monthly BBC science and technology magazine *Focus*, the mailbag brought several of them every day—missives from readers claiming to have cracked the mysteries of black holes, parallel universes, or the Big Bang; determined the origin of life; or unified the laws of particle physics. I'd reply, thanking them for their theories, and requesting that they send in the full supporting mathematics. I don't think any of them ever did.

That's the difference between the "theories" we bandy about in everyday parlance—our inklings and just-thought-of guesstimates—and the theories that are painstakingly constructed by scientists.

A theory in science is a logical creation. It reflects the most accurate experimental observations and the best understanding of how the world works. Yet a scientific theory doesn't necessarily represent absolute truth. It can only capture the state of our knowledge so far. There's every chance that a new piece of evidence will come to light that disproves the theory, and sends the theoreticians back to the drawing board.

One example of this is our view of the Solar System. In the second century CE, the Greek philosopher Ptolemy developed the theory that the Earth lies at the center of the Solar System—a sound explanation for the primitive astronomical observations of the day. But, in the early 17th century, the Italian astronomer Galileo began to survey the skies with the newly invented telescope. It permitted observations of the Solar System that were vastly superior to anything achieved with the naked eye.

Galileo's observations revealed details that fitted with a new theory, developed by the Polish astronomer Nicolaus Copernicus a hundred years earlier. Copernicus' theory painted the now-familiar picture of the Sun, not the Earth, sitting at the heart of our Solar System. Many observations since—including data from space probes—have confirmed the Sun-centered view of the Solar System.

Other casualties include the flat Earth theory, phlogiston theory—an early attempt to explain the origin of fire—and so-called intelligent design theory. Our theoretical understanding of pretty much every branch of modern science has evolved in this way, with old, defunct theories being replaced by new and improved ones.

The theories composing the body of scientific knowledge today cover everything from the origin of the Universe to the workings of the human mind. Over the pages that follow, the 50 greatest theories are laid out by some of the most talented science communicators. Each is summed up in a single, user-friendly passage encapsulating its essence. No jargon, no waffle—just concise, plain English.

The theories are organized into seven pillars of understanding. The first is The Macrocosm, and deals with the large-scale physics of the everyday world, such as the laws of motion, gravity, and electricity. The Microcosm turns our attention to the very small, looking at the quantum world of atoms and other subatomic particles of nature. The third pillar focuses on Human Evolution—how life, people, and facets such as intelligence and language all came to be. Mind & Body charts key theories in medicine— from psychoanalysis to gene therapy. In Planet Earth, we survey the great theories that have enabled scientists to grasp the inner workings of our planet and its climate. The Universe casts an eye further afield, taking stock of the origin, evolution, and ultimate fate of our cosmos—and others. The final pillar, The Knowledge, deals with branches of science concerned with the growth of science itself, such as Moore's Law for the constant improvement in the power of computers and Ockham's Razor— quite literally, the mother of all theories. The pillars also include profiles of some of the giants in these fields—summarizing the lives of intellectual heavyweights from Charles Darwin to Stephen Hawking.

This book serves a dual purpose. Its structured, piecemeal approach makes it an excellent reference to dip into as required—a mini-encyclopedia of theoretical science. On the other hand, read it cover to cover, and you'll have an excellent overview of how scientists today think the natural world works. So, if you're in a quandary about quantum theory or wrestling with relativity, or just curious about what exactly scientists have been up to all these years, then sit back in your favorite armchair and let our resident experts guide you through the greatest achievements of the human mind. But please—keep your pet theories to yourselves!

Relative values

The theory of relativity is probably one of the best-known scientific theories—but do we really understand it? It's all about how time, matter, energy, and space interact (apparently)—see page 30 for the half-minute explanation.

THE MACROCOSM

electric current is a flow of electrons (or other charged objects) from a negatively charged object to a positively charged one.

electromagnetic wave Another way of describing radiation, such as light and heat.

equation A mathematical notation used to show how measurable quantities relate to each other. $E=mc^2$ is an equation showing how the energy in an object (E) equals the object's mass (m) multiplied by the speed of light (c) squared (^{2}). (Squared means a number multiplied by itself once.)

field An area of space in which a force has an effect on matter. Examples include magnetic and gravitational fields.

kinetic energy The energy contained in a moving body that relates to its motion.

law A simple description of a pattern that has been observed in nature. Most laws are expressed as equations.

macrocosm The big picture—a model that reflects the functioning of a system on the largest of scales.

mass A measure of the quantity of matter in an object. "Mass" and "weight" are often used interchangeably, but weight is really a measure of the pull of gravity on the object. In everyday terms, the "mass" and "weight" of an object are effectively the same on Earth, but on the Moon, the same object's mass is unchanged, while its weight is reduced by 85 percent by the lower gravity.

matter The stuff of the Universe, which fills space and can be measured in some way.

oscillations Rhythmic movements which occur around a central, unchanging point in space.

particles Small units of matter. In physics on the tiniest length scales, a particle may be a minute building block within an atom, or a molecule of water, oxygen, or any other substance. Otherwise it may be a speck of dust, smoke, or sand among many.

potential energy The energy stored within an object that could be released and harnessed to do useful work. A boulder teetering on a hilltop has potential energy. If it is pushed down the hill, that potential will be converted to kinetic, or movement, energy.

perpendicular At a right angle—90 degrees—to something else. Walls are perpendicular to the floor—hopefully.

radiation A term sometimes used to describe the dangerous emissions from radioactive substances, but more correctly used to describe the transfer of photons—tiny packets of energy—through space. Light, heat, radio waves, as well as dangerous gamma rays, are all types of radiation, each carrying varying amounts of energy.

refraction This is when a beam of light, or other radiation, changes direction slightly as it is passed from one medium (e.g. the air) to another (e.g. water). The refraction is due to the differences in the speed of light in the two media. If a beam of light arrives at the interface between the two media at an angle, one side of the beam will change speed before the other side. As a result the whole beam changes direction slightly.

speed of light The speed at which radiation travels, and the speed limit of the Universe. The speed of light in a vacuum is 186,282 miles per second (299,792 km/s). Nothing can travel faster than this.

subatomic Smaller than an atom.

PRINCIPLE OF LEACTION

the 30-second theory

This says, essentially, that things happen in the way that requires least effort. So, a beam of light will travel in a straight line because that is the shortest path between two points. If you drop a ball, it will travel toward the center of the Earth. No one is quite sure who came up with the principle of least action, but your everyday experience would probably lead you to come up with it if you thought about it for a bit. In the 18th century, though, this was a big deal. Some of the greatest names in mathematics, such as Leonhard Euler, Pierre de Fermat, Gottfried Leibnitz, and Voltaire were involved in the argument over who came up with the idea first. It was important to make these kinds of statements at the time, because they led to the formation of the equations that describe how things move when acted on by forces. They also led to the concepts of potential and kinetic energy.

RELATED THEORIES
see also
UNIFICATION
page 50
OCKHAM'S RAZOR
page 142

3-SECOND BIOGRAPHIES
LEONHARD EULER
1707–1783
PIERRE DE FERMAT
1601–1665
GOTTFRIED LEIBNITZ
1646–1716
VOLTAIRE
1694–1778

30-SECOND TEXT
Michael Brooks

3-SECOND THRASH
At the core of modern physics is the notion that "Nature is thrifty in all its actions ..."

3-MINUTE THOUGHT
Quantum theory, which describes how things work on the subatomic scale, seems to be the one area where the principle of least action does not apply. Quantum objects can be in two states at once, and can take multiple paths when traveling from one place to another. Richard Feynman went so far as to suggest that a quantum particle will simultaneously take every possible path when making a journey!

As theories go, the principle of least action is just common sense: Natural motion always takes the easiest and shortest route.

Macrocosm

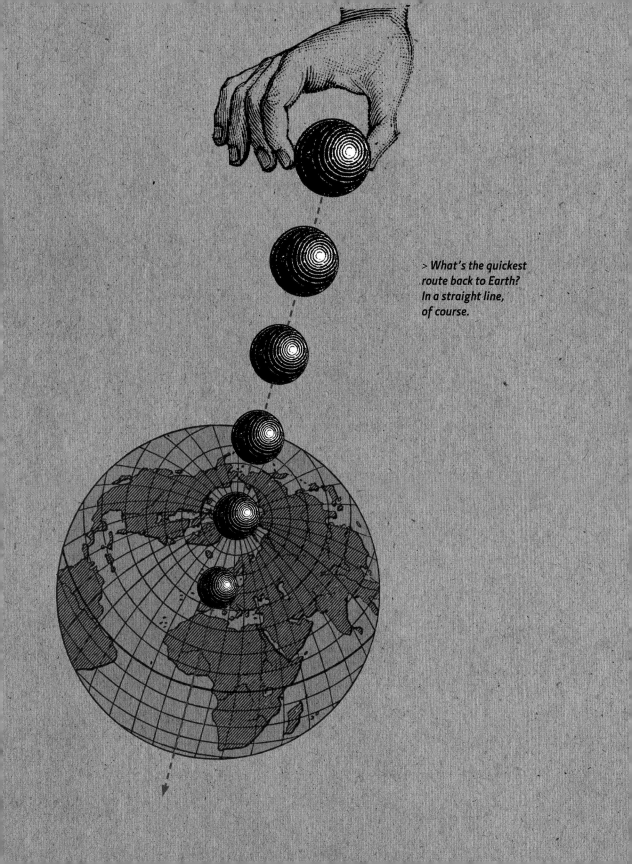

> *What's the quickest route back to Earth? In a straight line, of course.*

LAWS OF MOTION

the 30-second theory

When Isaac Newton sat down

and thought about how things move, he worked out three laws that are now so familiar they seem like common sense. First, he said that objects have "inertia," which is a measure of resistance to changes in their motion. Inertia means that things remain still until you give them a push. Similarly, objects that are moving keep moving unless something stops or pushes on them. Second, the mass of the object determines what effect a particular push will have on the motion (or lack of it). The third law, which is the most famous, feels slightly different. It says that every action has an equal and opposite reaction. If I push you, I feel an equal push in return. This is the principle by which space rockets and jet engines work: When they push out an exhaust gas from the nozzle at the rear, the engines get a push forward. This is why you should be careful when you step off a boat. To move yourself forward, you inevitably move the boat backward. If you don't take that into account, you can end up taking a swim!

3-SECOND THRASH
Newton formulated the basic description of how things move and effectively invented rocket science.

3-MINUTE THOUGHT
Newton's laws are simple but powerfully accurate. They are not accurate enough, however, to describe what happens when things are moving at close to the speed of light, or in strong gravitational fields. In such instances, Einstein's theory of relativity takes over and provides our ultimate laws of motion.

RELATED THEORIES
see also
UNIVERSAL GRAVITATION THEORY
page 20
THEORY OF RELATIVITY
page 30
UNIFICATION
page 50

3-SECOND BIOGRAPHY
ISAAC NEWTON
1643–1727

30-SECOND TEXT
Michael Brooks

The laws of motion are all you need to describe how everyday objects move—from soccer balls to space stations. Newton gave us the means to plan a journey to the Moon—it just took 300 years to invent the rockets needed to do it.

> *Fly me to the Moon —using Newton's laws of motion.*

UNIVERSAL GRAVITATION THEORY

the 30-second theory

This description of one of the fundamental forces of nature is among the greatest achievements in science. Isaac Newton came up with it in 1687 as part of his masterful *Principia Mathematica*, a three-volume description of mathematics. Universal gravitation theory says that there is a mutual attraction between anything that has mass—anything made of normal matter, that is. That attraction depends on the two masses involved, the distance between them, and a constant known as the gravitational constant. One of the central insights of the theory was that the gravitational force follows an "inverse square law." This means the attraction between the two objects diminishes as the square of the distance between them. Newton's formulation of the law was so accurate that it immediately explained the motion of the planets, creating an easy way to predict their movements relative to each other and the Sun. It has also enabled us to send rockets into space. After Einstein came up with the theory of relativity and used it to explain some small anomalies in the planetary orbits, it was realized that Newton's law was not quite the final word on gravity. However, it is almost universally accurate when applied to the gravitational attractions we encounter in everyday life.

RELATED THEORIES
see also
LAWS OF MOTION
page 18
THEORY OF RELATIVITY
page 30
QUANTUM FIELD THEORY
page 46
UNIFICATION
page 50

3-SECOND BIOGRAPHY
ISAAC NEWTON
1643–1727

30-SECOND TEXT
Michael Brooks

3-SECOND THRASH
What goes up must come down—and it will do so, just as Newton said it should.

3-MINUTE THOUGHT
Some ideas in modern physics suggest that Newton's law of gravitation may need adjusting to consider things separated by less than a millimeter, or more than the diameter of the Solar System. What's more, no one has a good explanation for why things with mass attract each other in the first place, why gravity is much weaker than the other forces of nature, or for the true value of the gravitational constant, which is the least-accurately-measured constant of physics.

Large or small, everything comes down to Earth with the same bump.

Macrocosm

$$F = G \frac{m_1 \times m_2}{r^2}$$

> The acceleration force due to gravity is the same for a massive elephant as it is for a tiny pea—but get out of the way of falling elephants!